Intro to Science

Student Diary

Intro to Science Student Diary

Third Edition (First Printing 2024)

ISBN # 978-1-953490-24-7

Printed in the USA for worldwide distribution

For more copies write to:
Elemental Science
PO Box 79
Niceville, FL 32588
support@elementalscience.com

Copyright Policy

Classical SCIENCE

A Quick Welcome from the Author

Dear Student,

Welcome to *Intro to Science!* This workbook will serve as a scrapbook of sorts for you to share what you learned about science. You will be learning about the basics of science this year.

Each week you and your teacher will do the following:

- **Read** the assigned pages together. Your teacher will then ask you a few questions as you discuss what was read. Be sure to share what you found interesting.
- **Do** the weekly demonstration with your teacher. This is the super fun part of science, plus you get to exercise your observation muscles. Be sure to pay close attention and help out when your teachers ask you to do so.
- **Write** down what you have learned and seen. Your teacher may help you with the actual writing, but be sure to record the facts that you want to remember.

Your teacher has the tools to add in more each week, things like memory work, library books, and extra activities. Be sure to let them know if you want to dig deeper into a topic.

And, if you have a question or want to share your work with me, please have your teacher send us an email (support@elementalscience.com) or by tagging us (@elementalscience) in a photo you share online. I would love to see what you have learned this year!

I hope that you enjoy learning about science this year!

Paige Hudson

Table of Contents

A Peek Inside Your Student Diary

This student diary is a place to keep a scrapbook of what you learn this year.

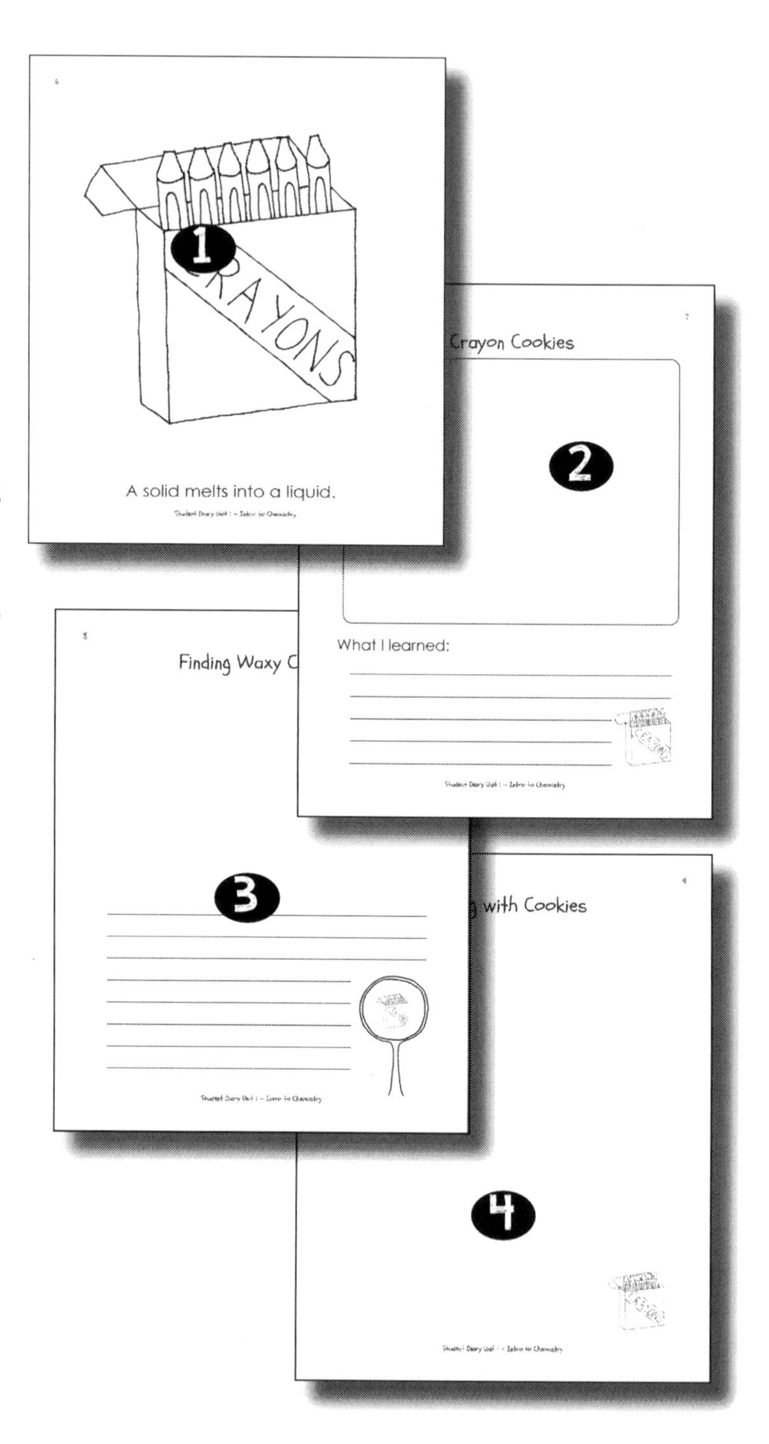

1. Main Idea Coloring Pages

These pages have the main idea at the bottom of the page. You can color the picture and have your teacher help you read the sentence to you.

2. Simple Demonstration Sheets

These sheets have space for you to document what you did during the scientific demonstration. Your teacher can help you write down what you learned on the lines provided and add pictures in the boxes provided.

3. Nature Journal Sheets

These sheets let you record what you learned from your nature study time. Again, your teacher can help you write down what you learned on the lines provided and add pictures in the boxes provided.

4. Art Pages

These pages give you space to draw a picture or paste in a picture of the craft project you made.

And now that you know what is in your diary, let's dig in!

Intro to Science

Unit 1: Chemistry Diary

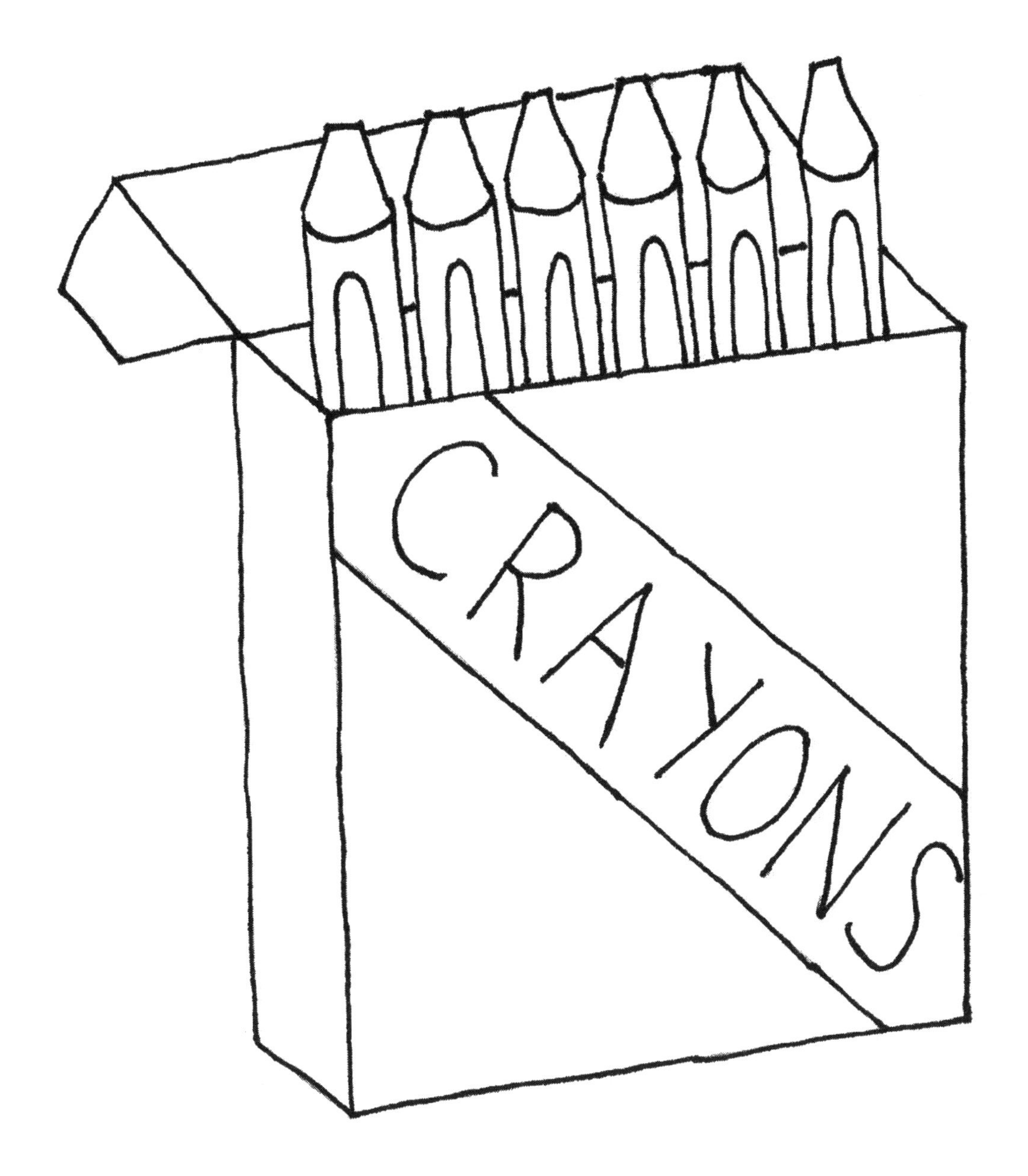

A solid melts into a liquid.

Crayon Muffins

What I learned:

Finding Waxy Coatings

Coloring with Crayon Muffins

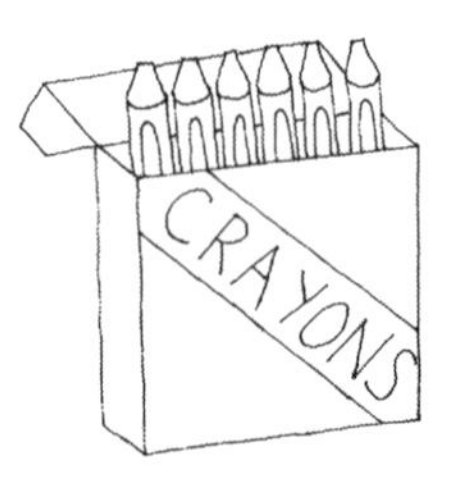

Thick paint

Thick paint after adding water

Adding water to a mixture will make it thinner or weaker.

Dilution Chemistry

Juice to Water Ratio	Rating
All Juice (1 cup Juice)	
Mostly Juice, Some Water (3/4 cup Juice, 1/4 cup Water)	
Half Juice, Half Water (1/2 cup Juice, 1/2 cup Water)	
Some Juice, Mostly Water (1/4 cup Juice, 3/4 cup Water)	
All Water (1 cup Water)	

What I learned:

Muddy Mixtures

Diluted Art

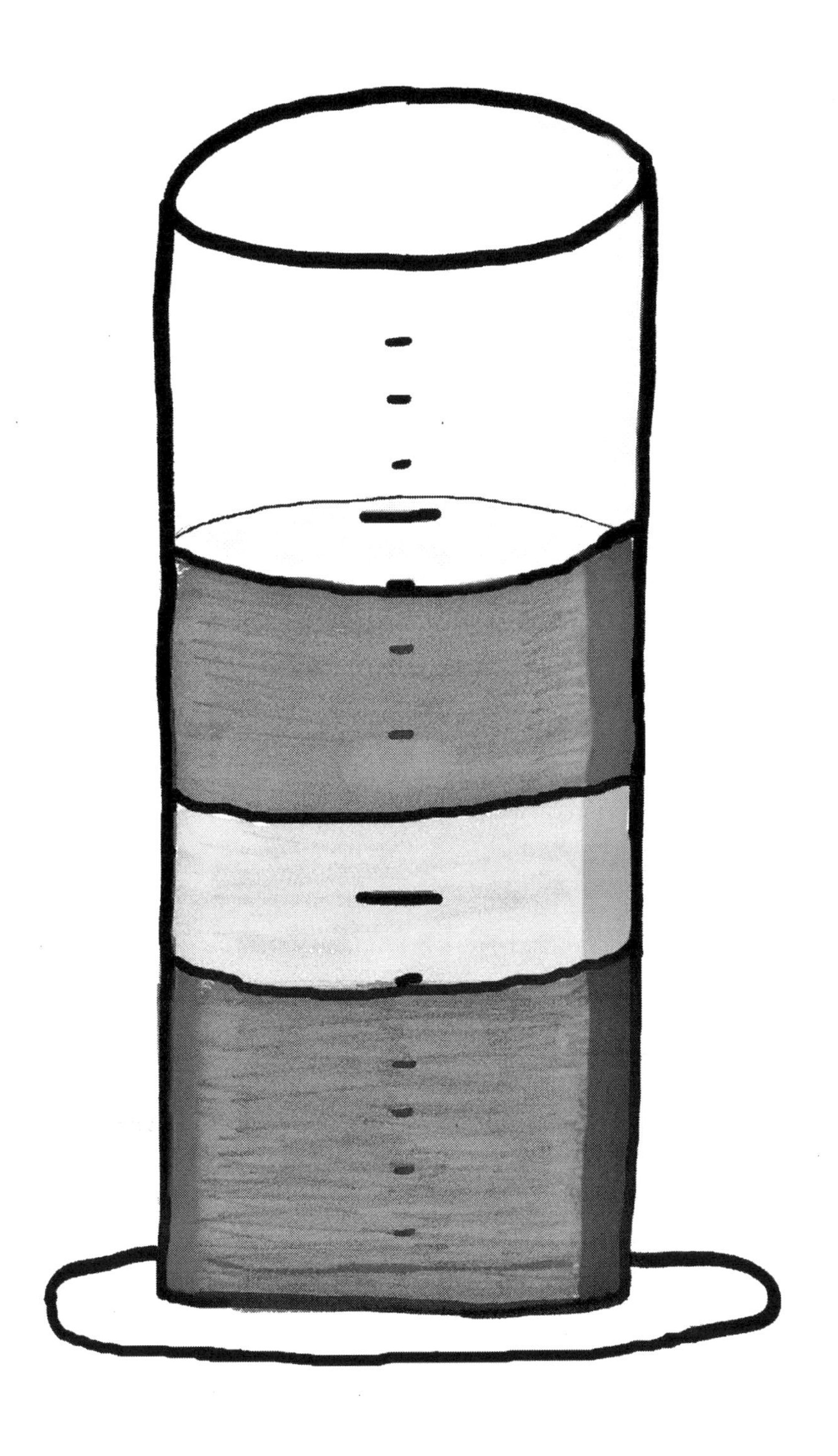

Oil is less dense than water.

Floating

Object	Sinks	Floats

Density in Nature

Object	Sinks	Floats

What I learned:

____

____

____

____

Marbled Paper

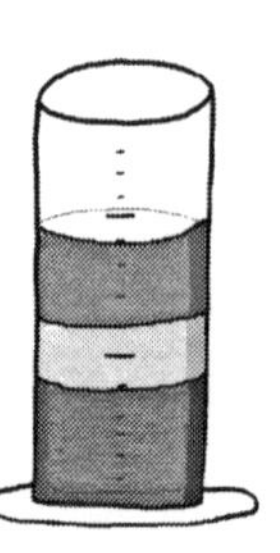

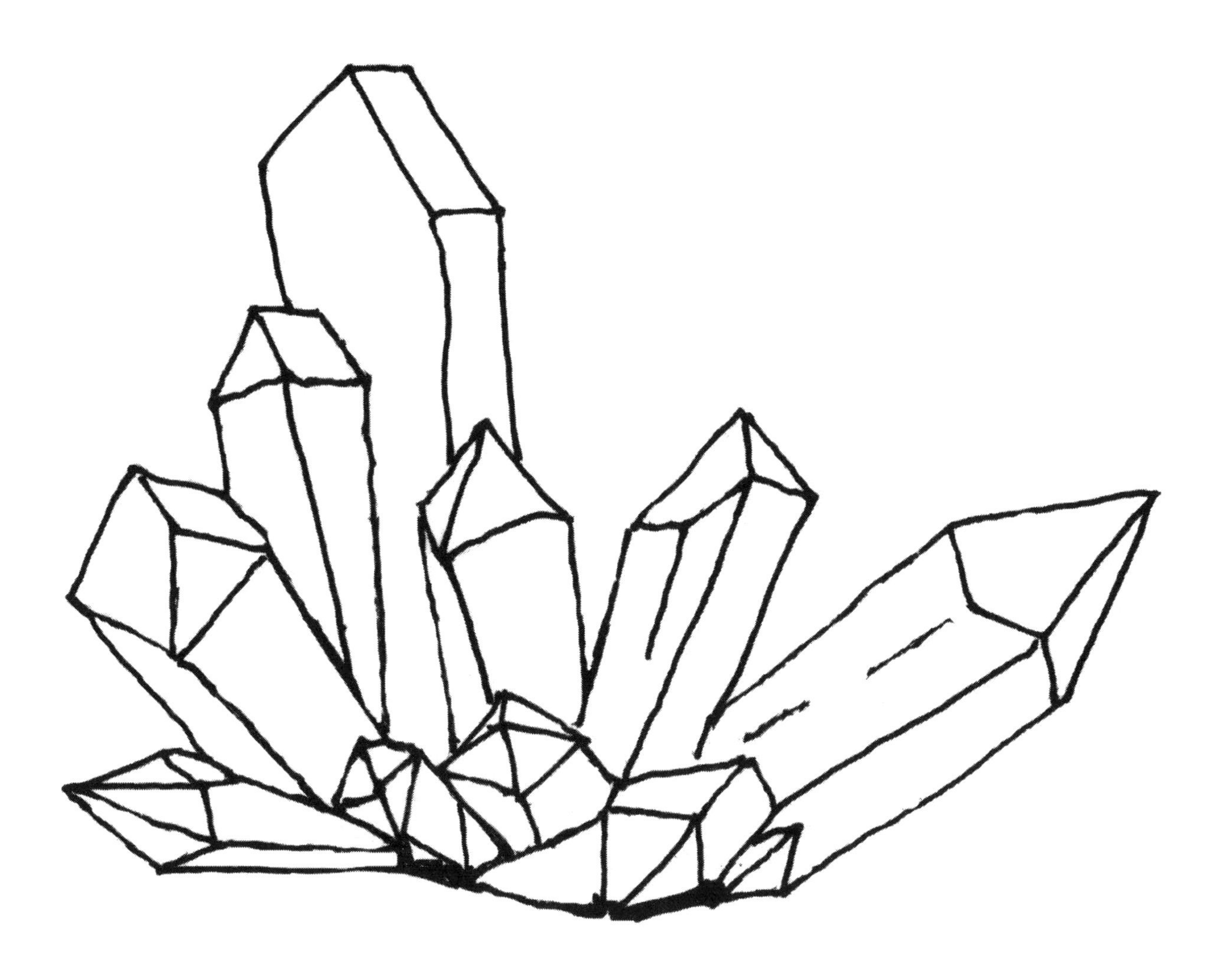

Crystals are made up of minerals found in the earth.

Crystals

What I learned:

Quartz

Painting with Crystals

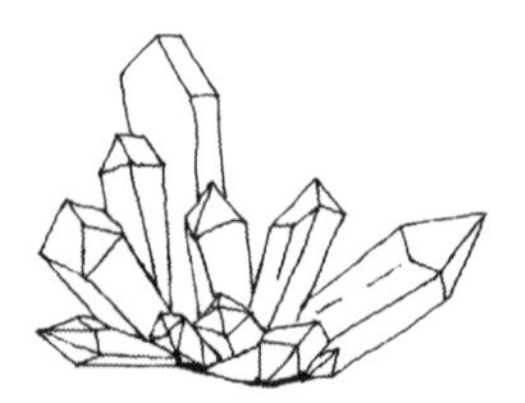

Our two colors

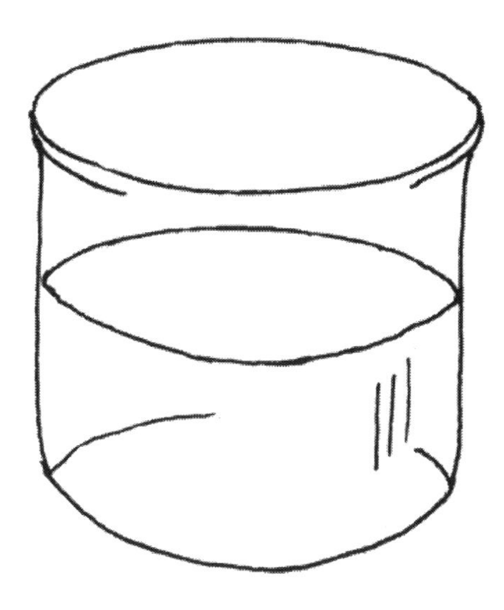

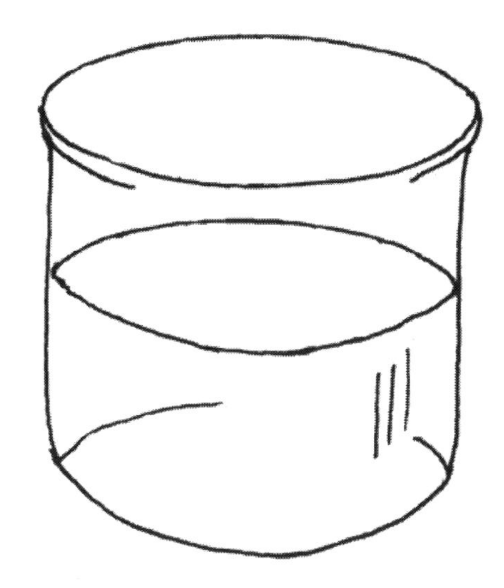

What happened when we mixed the colors

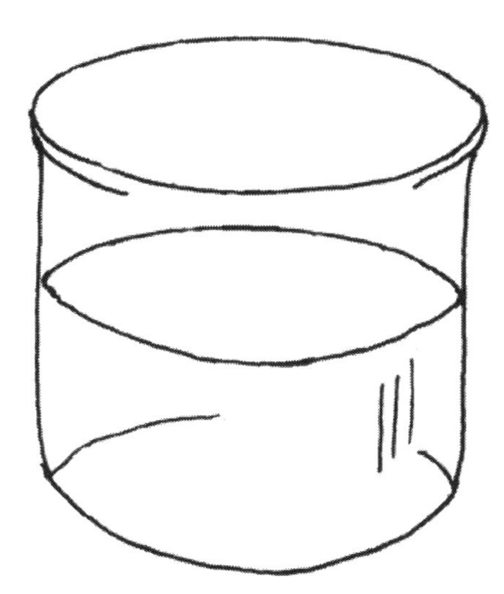

Two colors can be mixed to make a new color.

Color Mixing

What I learned:

Colors in Nature

Color Painting

When water freezes,
it changes into ice.

Freezing

Liquid	Did it freeze?
	Yes No
	Yes No
	Yes No
	Yes No
	Yes No
	Yes No
	Yes No

Weather Observation

Ice Painting

Intro to Science

Unit 2: Physics Diary

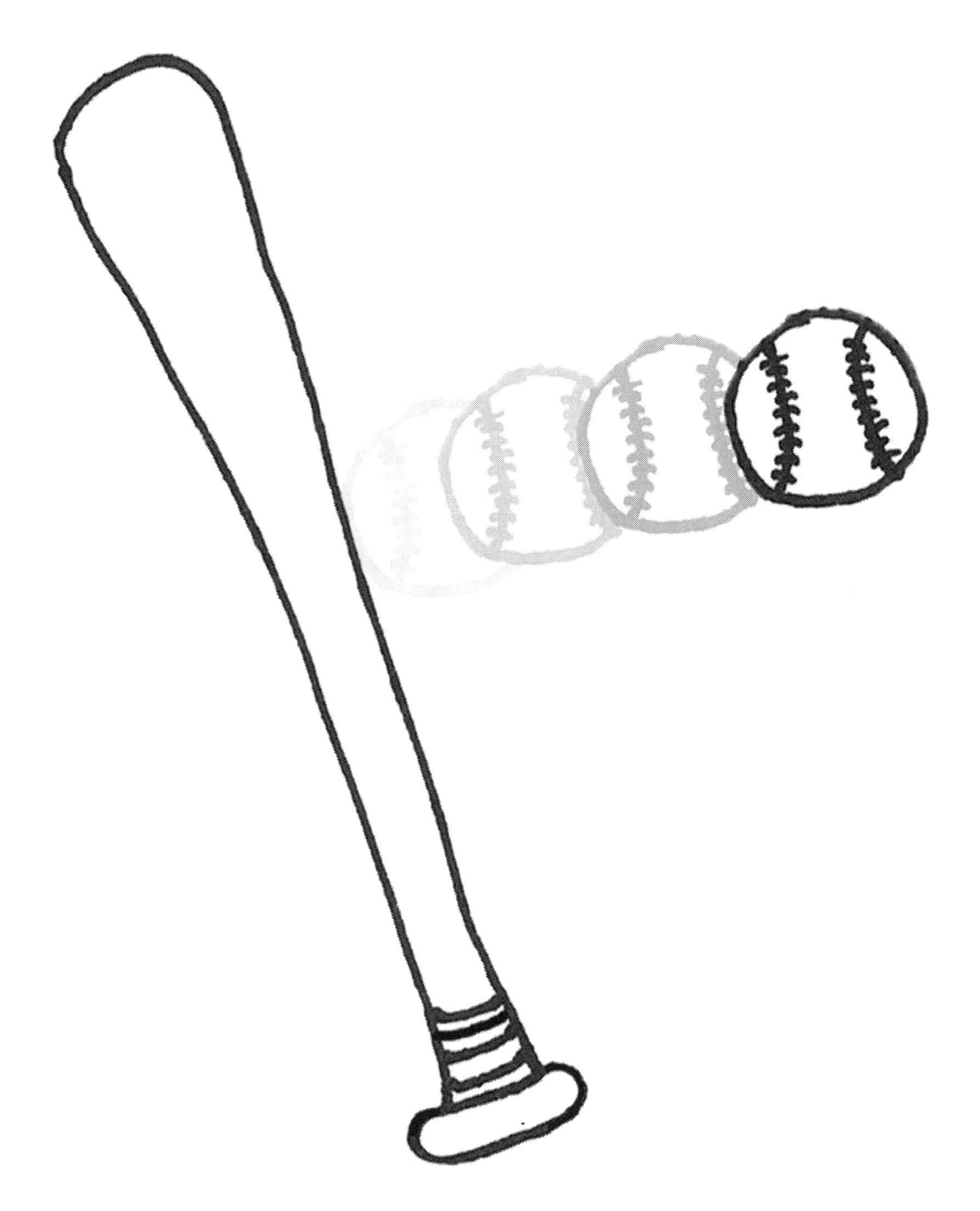

A force is a push or a pull that can cause motion or slow it down.

Push and Pull

	What the Car Did
On the Table	
After a Push	
After a Pull	

Weather Observation

Motion Painting

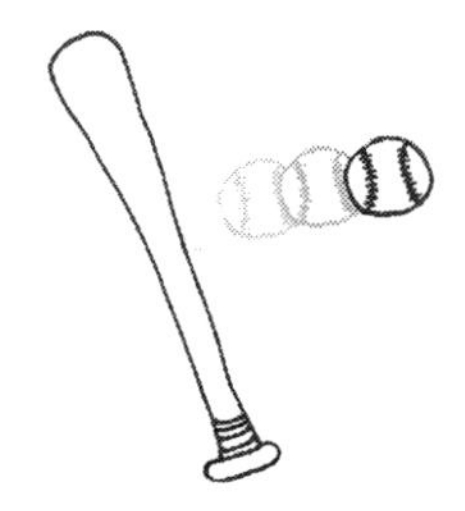

Gravity is the force that pulls all things to the ground.

Gravity Pull

What I learned:

Apple Tree

Gravity Drops

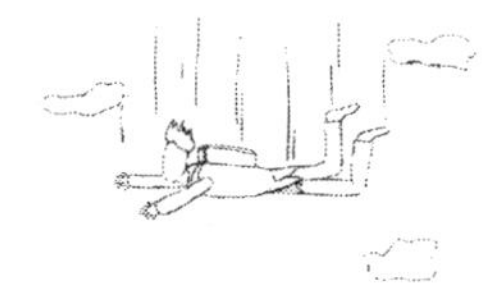

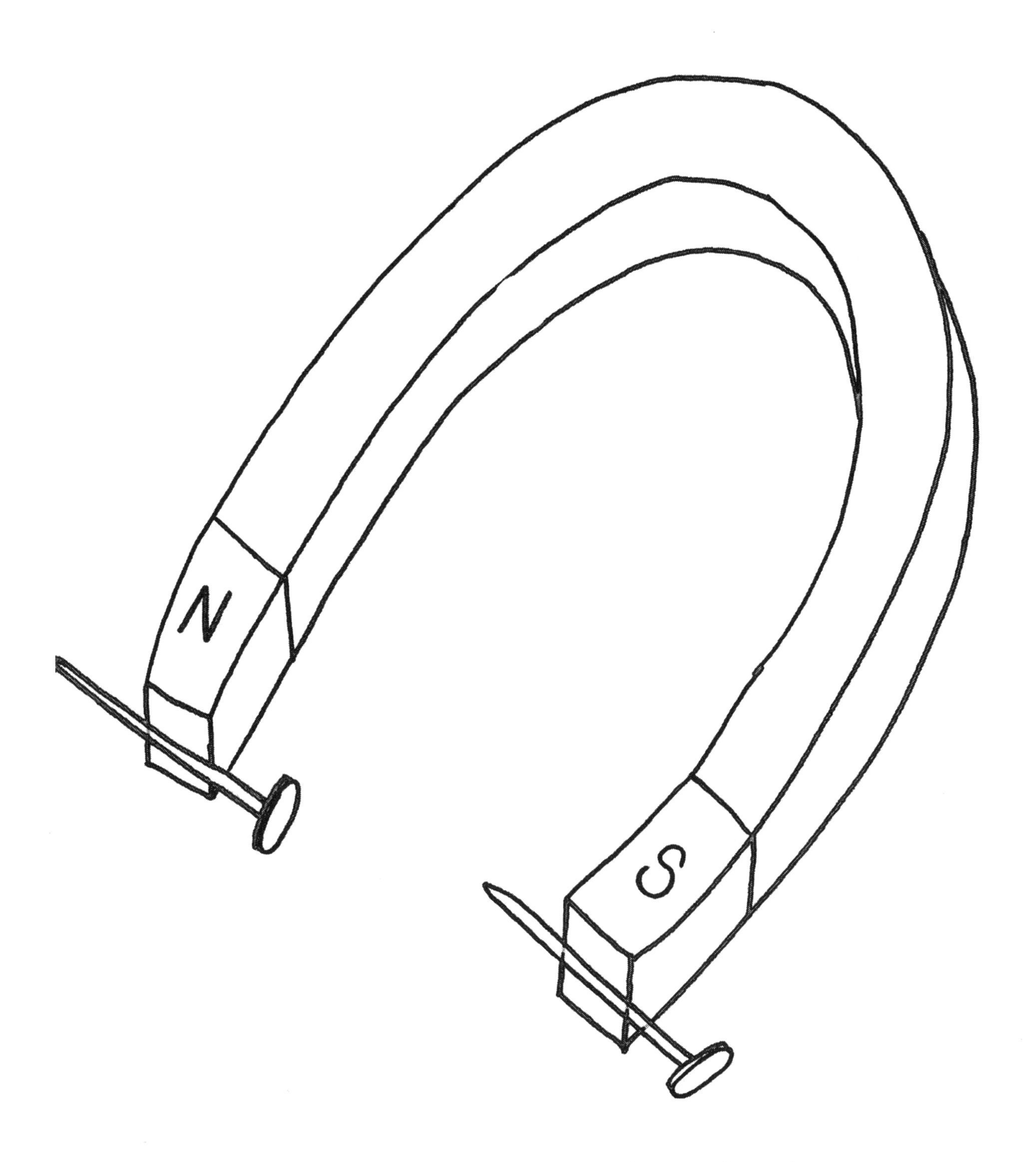

Magnets are attracted to certain kinds of metal.

Magnetic Attraction

Object	Was it attracted to the magnet?
	Yes No
	Yes No
	Yes No
	Yes No
	Yes No
	Yes No
	Yes No

Magnetism in Nature

Object	Was it attracted to the magnet?
	Yes No
	Yes No
	Yes No
	Yes No
	Yes No

What I learned:

____

____

____

____

Painting with Magnets

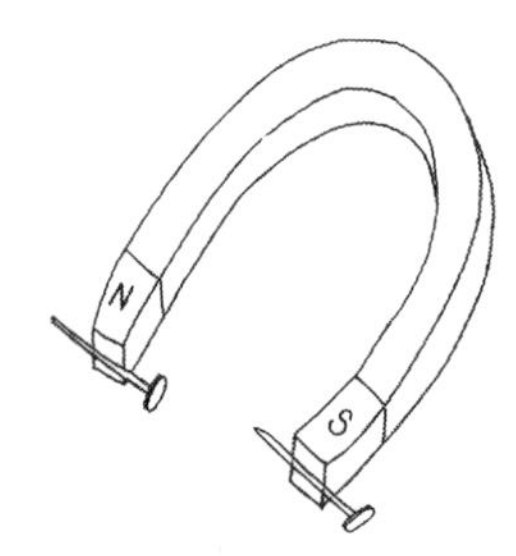

A ramp is called an inclined plane.

Ball Ramp

	Distance from the ramp
Marble	
Bouncy Ball	

Ramps in Nature

Ramp Painting

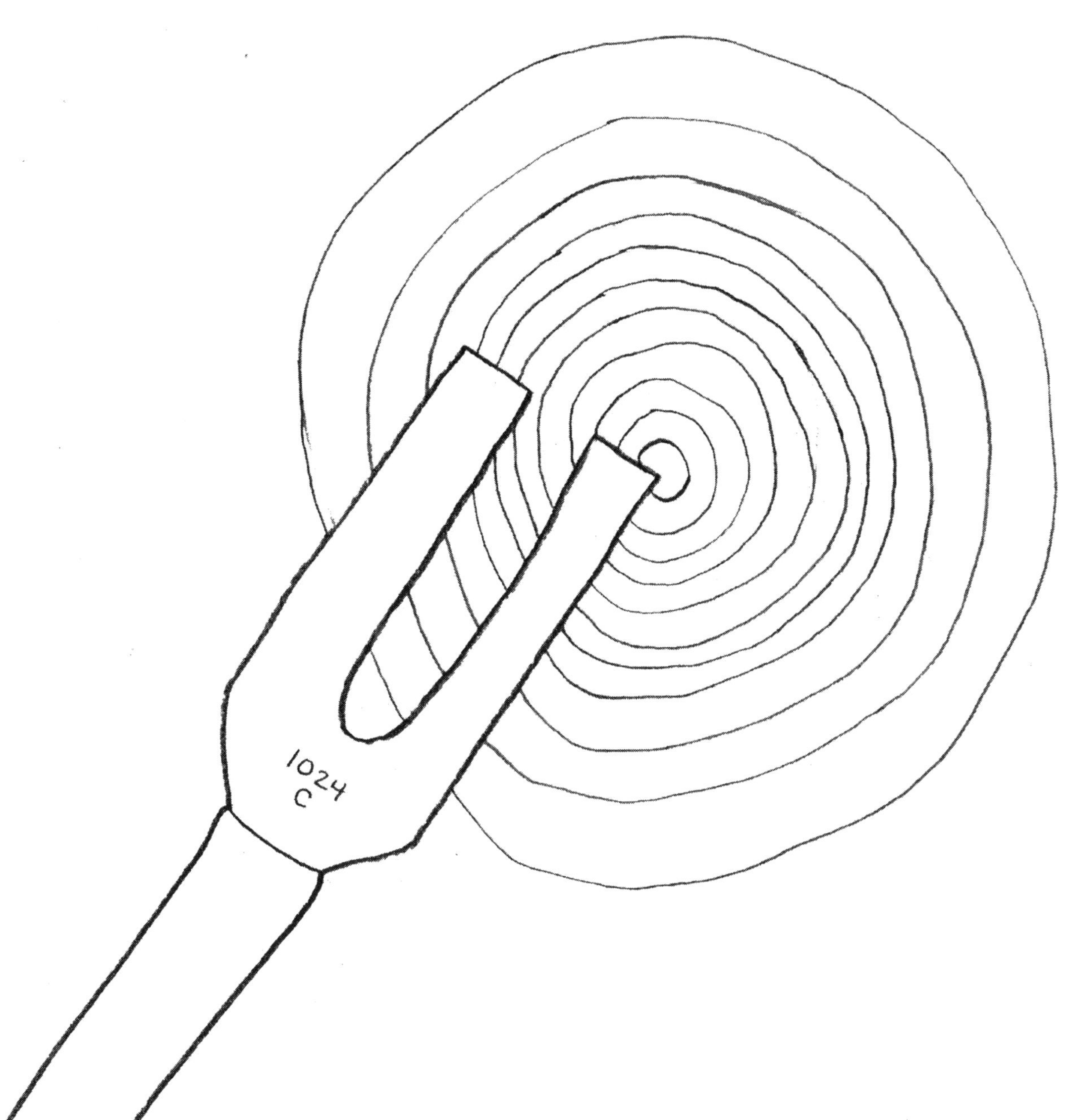

Sound waves are vibrations that can travel through the air.

My Tonoscope

What I learned:

Bird Calls

Plate Shakers

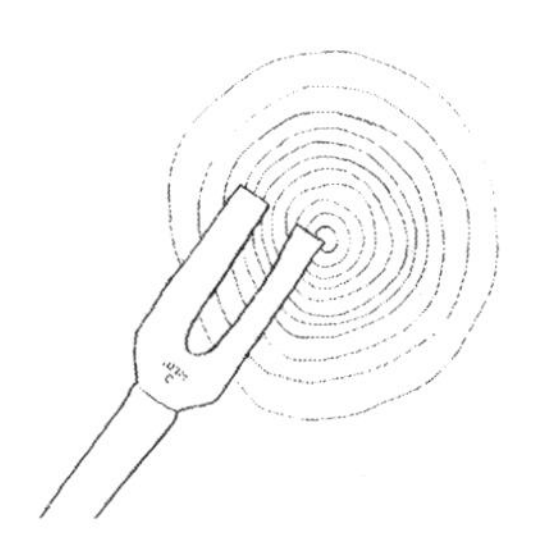

Light is the energy that helps us to see.

Shining Rainbows

What I learned:

The Sun

Reflection Collage

Intro to Science

Unit 3: Geology Diary

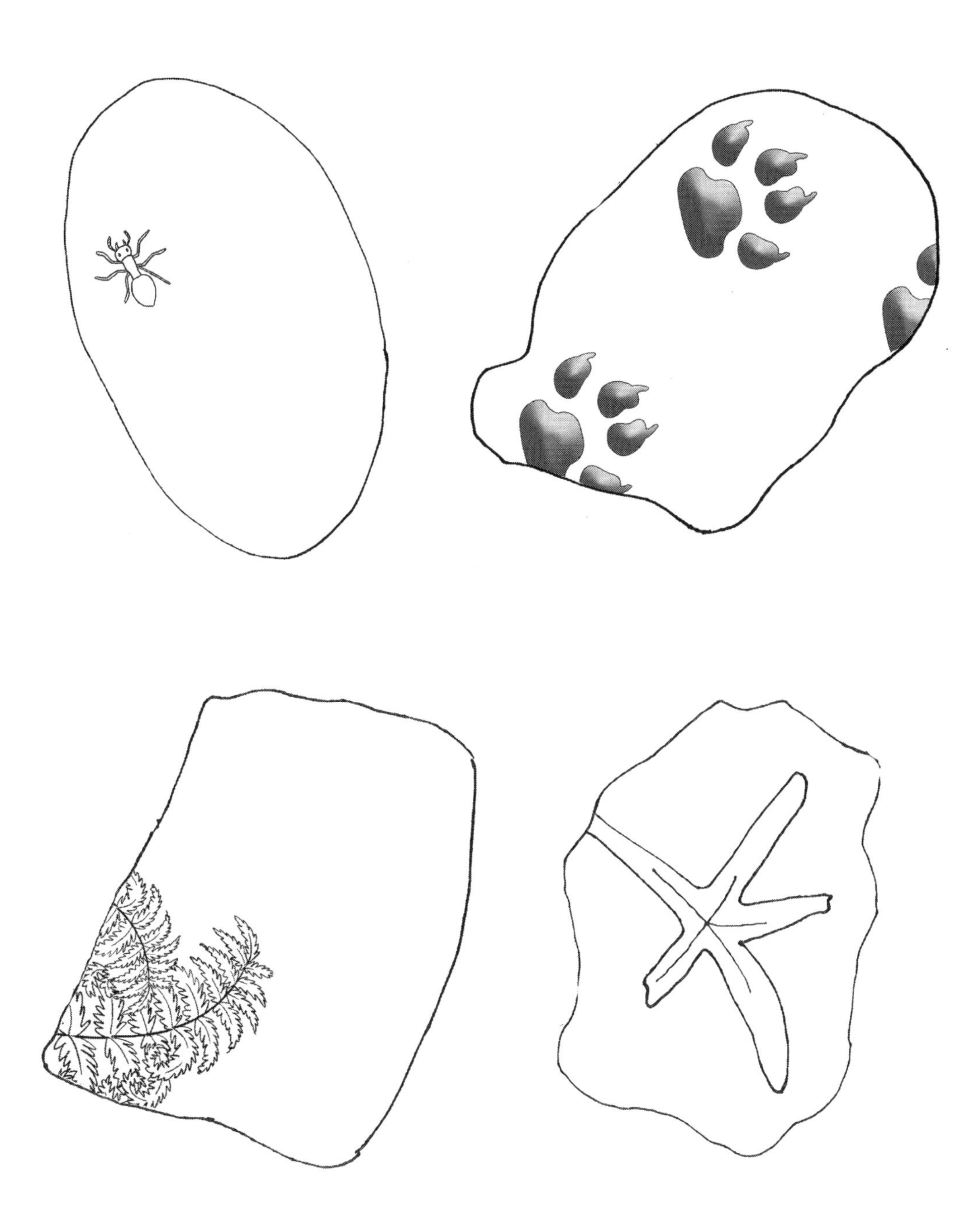

Fossils are imprints of
long-gone plants or animals.

Impression Fossils

What I learned:

Fossil Find

Fossil Prints

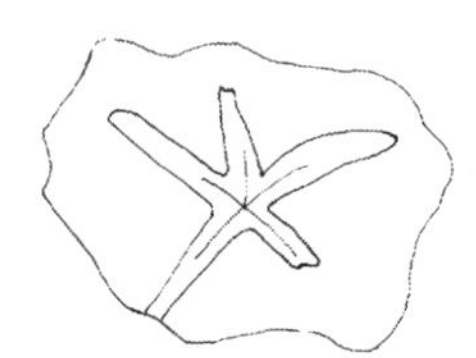

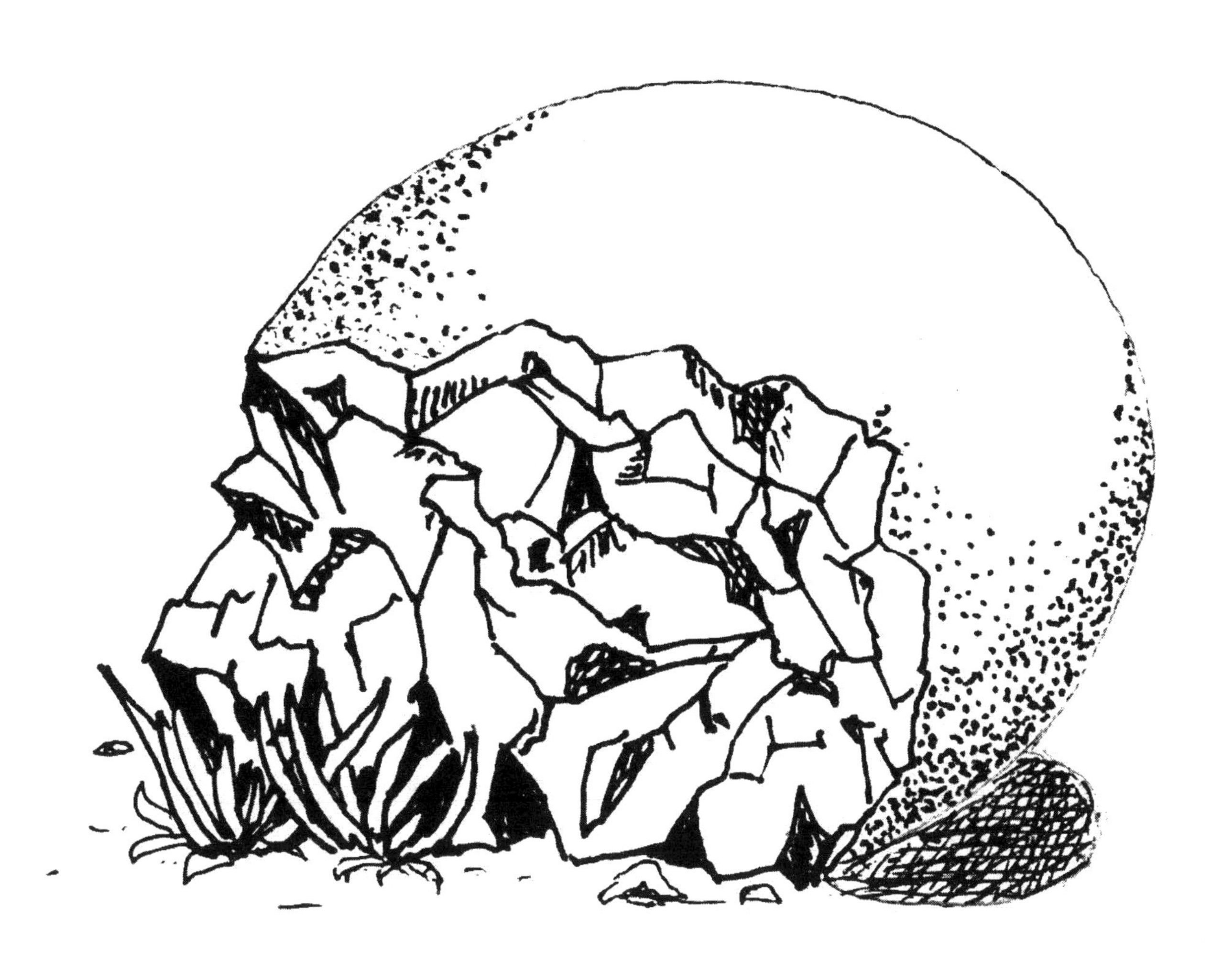

There are many different types of rocks.

Rocky Observations

	My Observations
How does the rock smell?	
How does the rock feel?	
What do you see on the surface of the rock?	
What color is the rock?	
What happens when I hit it with a hammer?	
What happens when you use the rock to mark on a paper?	

Rock Hunt

Painting Rocks

Metamorphic rocks are rocks that have changed.

Changed Rock

What I learned:

Metamorphic Rock Hunt

Metamorphic Art

Volcanoes explode hot, sticky rock from inside the Earth.

Toothpaste Volcano

What I learned:

Igneous Rock

My Volcano

Sedimentary rock is made from layers of sand, mud, or pebbles.

Sediment Jar

Before

After

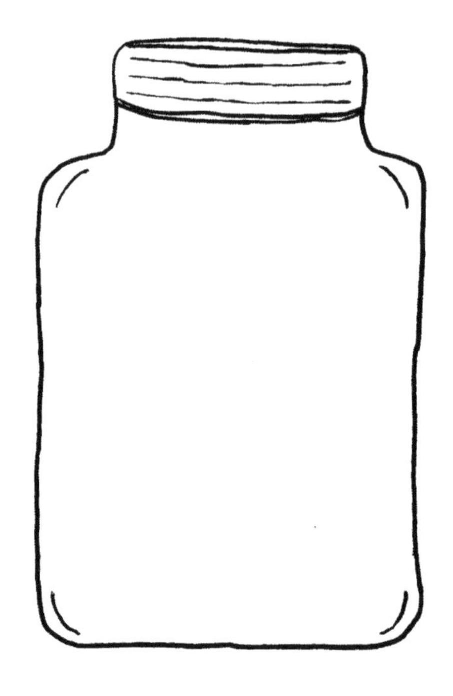

What I learned:

Sedimentary Rock

Sand Painting

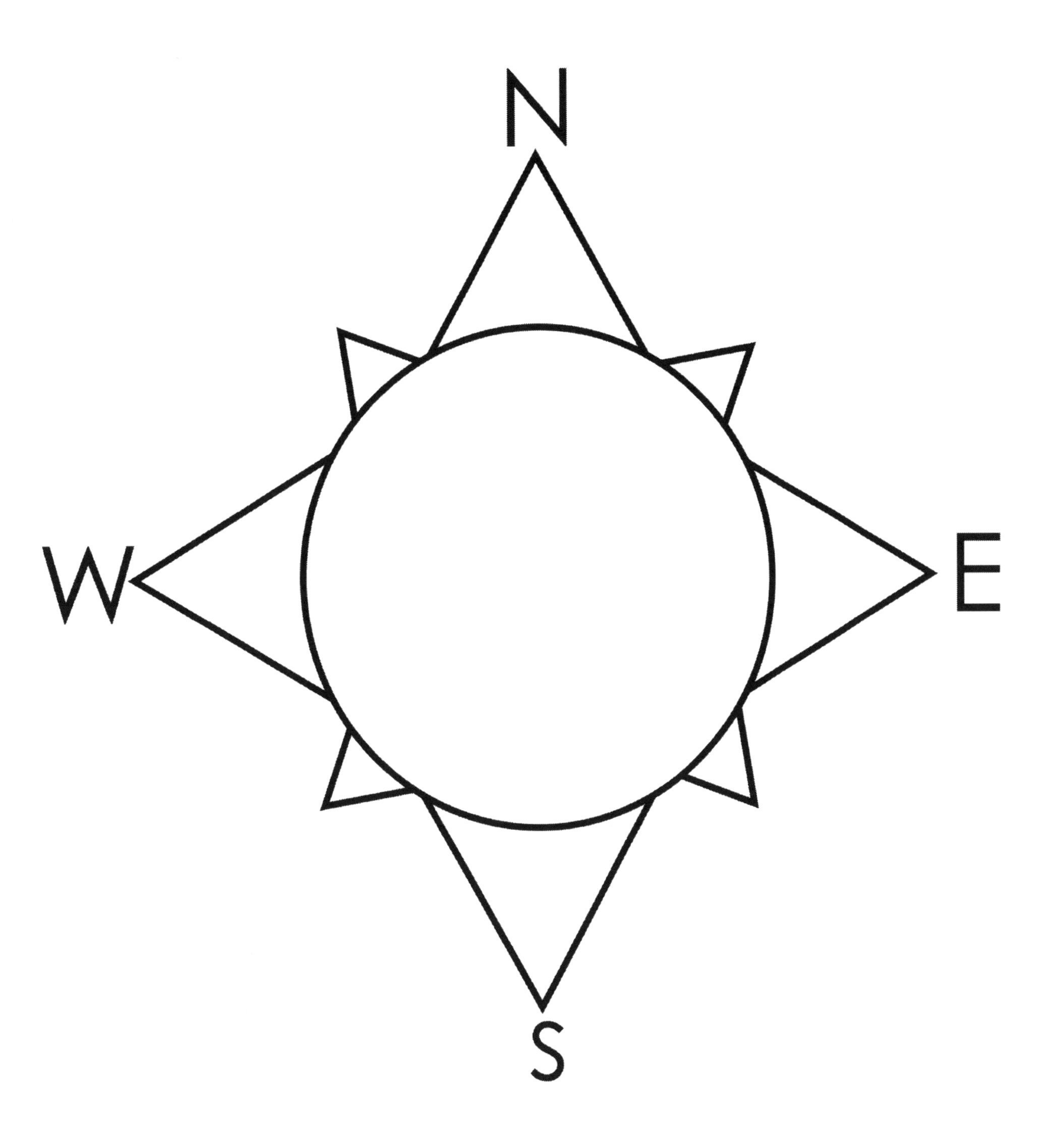

A compass shows us north, south, east, and west.

Treasure Hunt

What I learned:

N
W
E
S

Nature Map

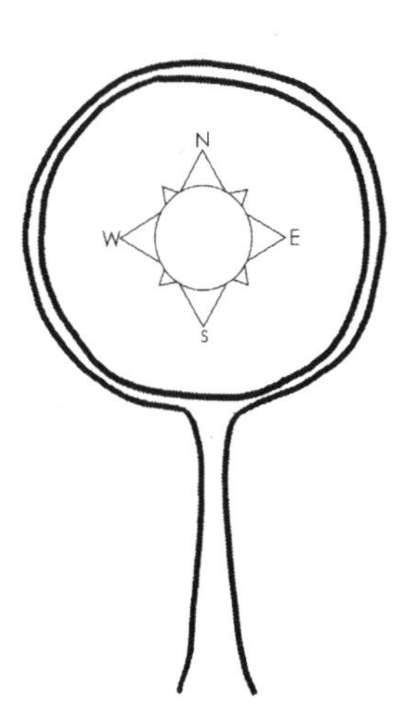

Room Map

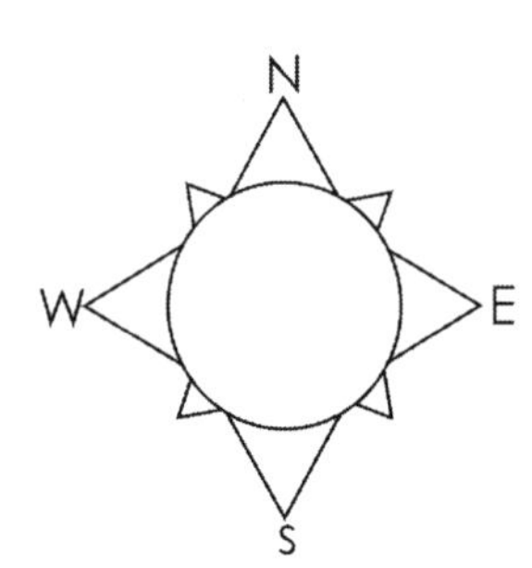

Intro to Science

Unit 4: Meteorology Diary

The energy from the
sun heats our earth.

Solar Melt

Food	What happened to it in the sun
Marshmallow	
Chocolate	
Butter	

What I learned:

____

____

____

____

____

Sunny Observations

Tissue Paper Sun

The water cycle shows the movement of water on the earth.

Water Cycle in a Jar

What I learned:

Dewy Observations

Raindrop Painting

Spring, summer, fall,
and winter are all seasons.

Weather Watch

Day	Weather
1	
2	
3	
4	
5	

Seasonal Tree Study

Seasons Collage

When air moves it causes wind.

Capture Wind

What I learned:

Wind

Draw a Storm

Tornadoes are funnels of spinning wind.

Tornado in a Jar

What I learned:

Tornadoes

Swirling Art

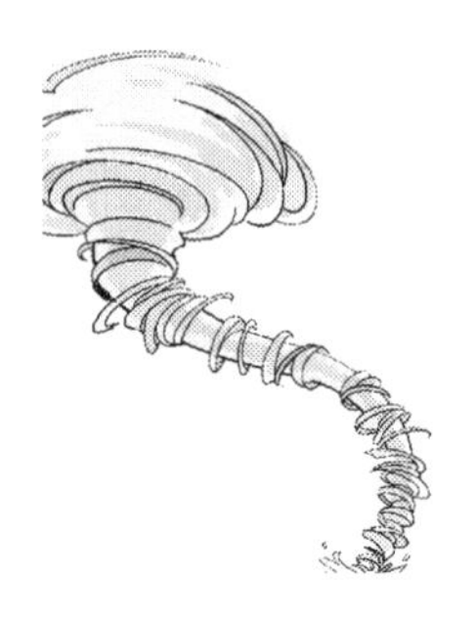

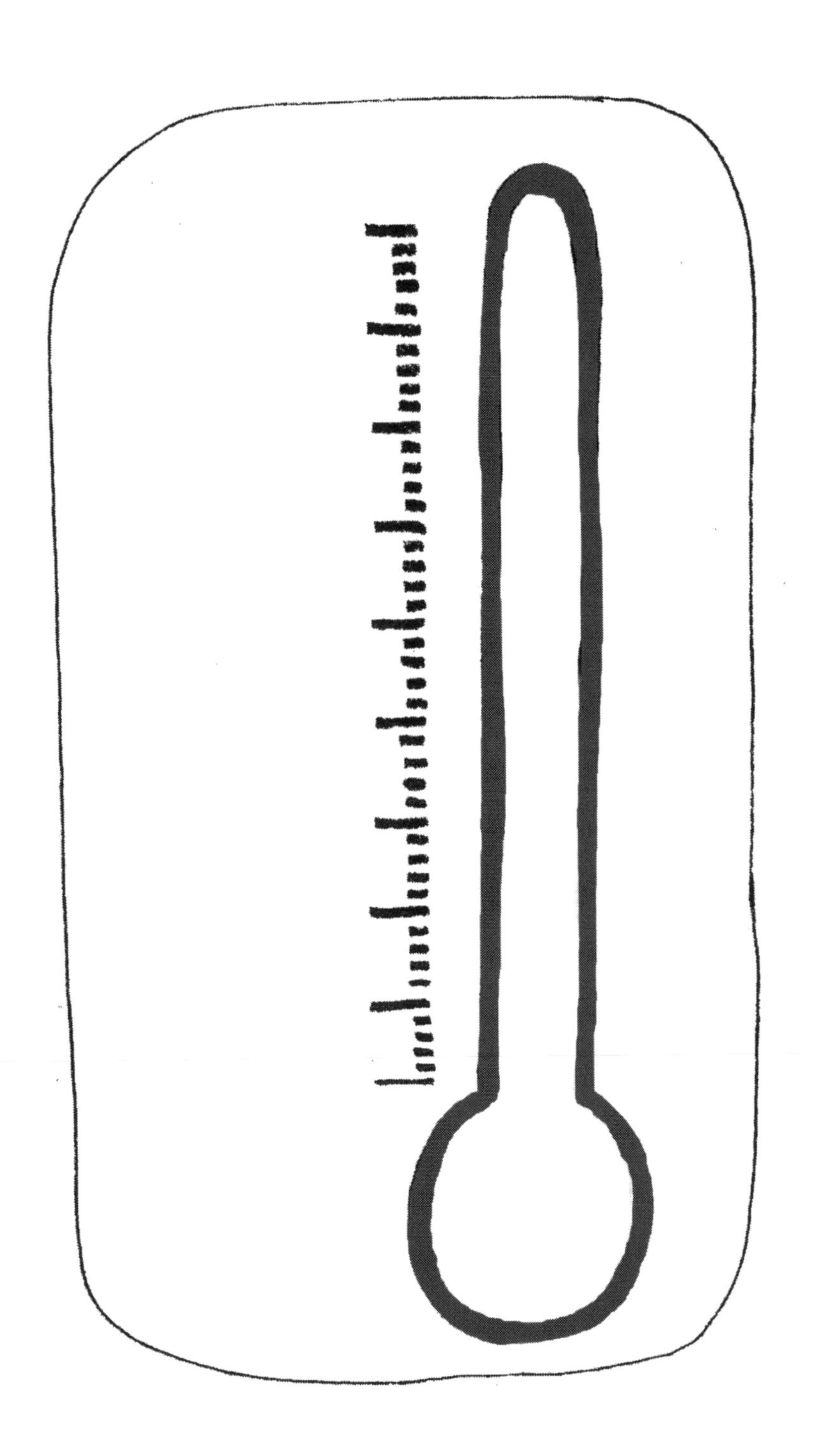

A thermometer tells us whether it is hot or cold.

Hot and Cold

	Cups
Temperature of Water at Start	Hot Water:
	Cold Water:
One Minute	Hot Water:
	Cold Water:
Two Minutes	Hot Water:
	Cold Water:
Three Minutes	Hot Water:
	Cold Water:

Temperature and Thermometers

Temperature Collage

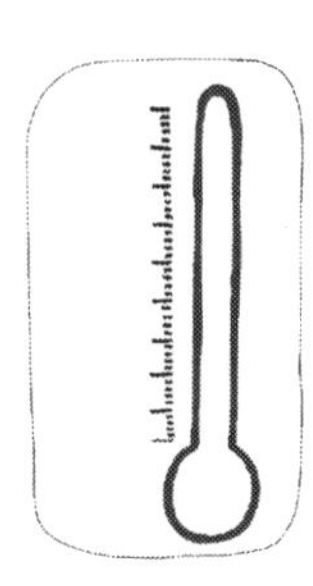

Intro to Science

Unit 5: Botany Diary

Plants grow toward the light.

Plant Growth

At the beginning of the week	At the end of the week

What I learned:

Plants

Mosaic Plant

Flowers have the parts of
a plant needed to make a seed.

Dissecting a Flower

What I learned:

Flowers

Field of Flowers

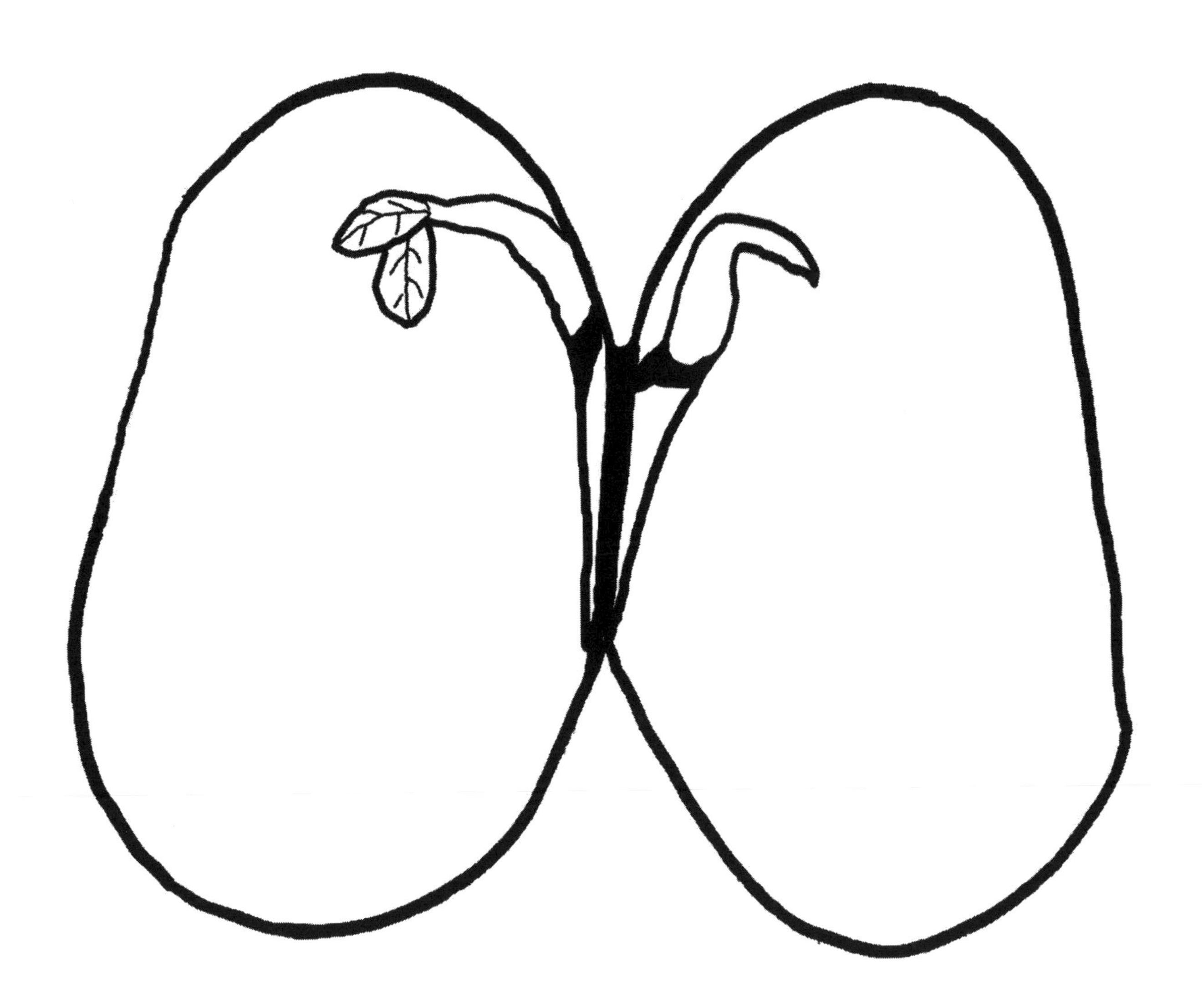

Seeds contain tiny baby plants.

Always Up

What I learned:

Seeds

Seed Prints

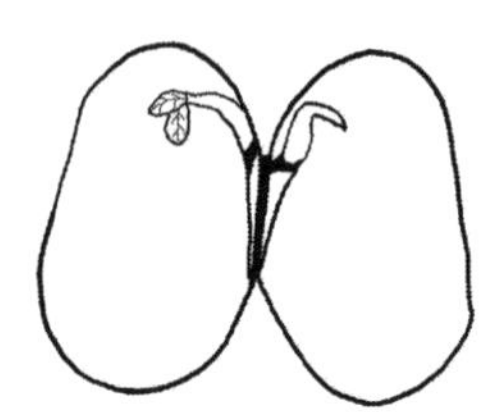

Leaves help the plant make food.

Leaf Cover-up

In the beginning

After 3 to 4 days

Leaves

Leaf Rubbings

The stem of a plant acts as its highway.

Thirsty Stems

Day 1

Day 3

What I learned:

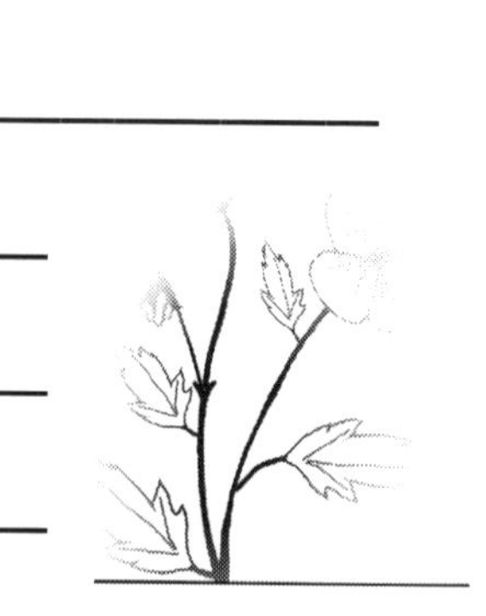

Oak Tree

Blowing Stems

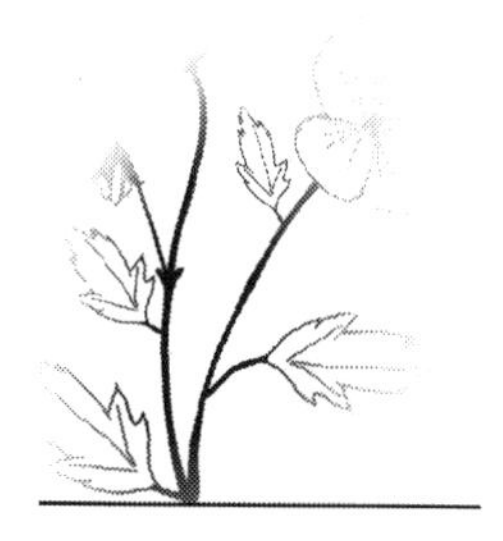

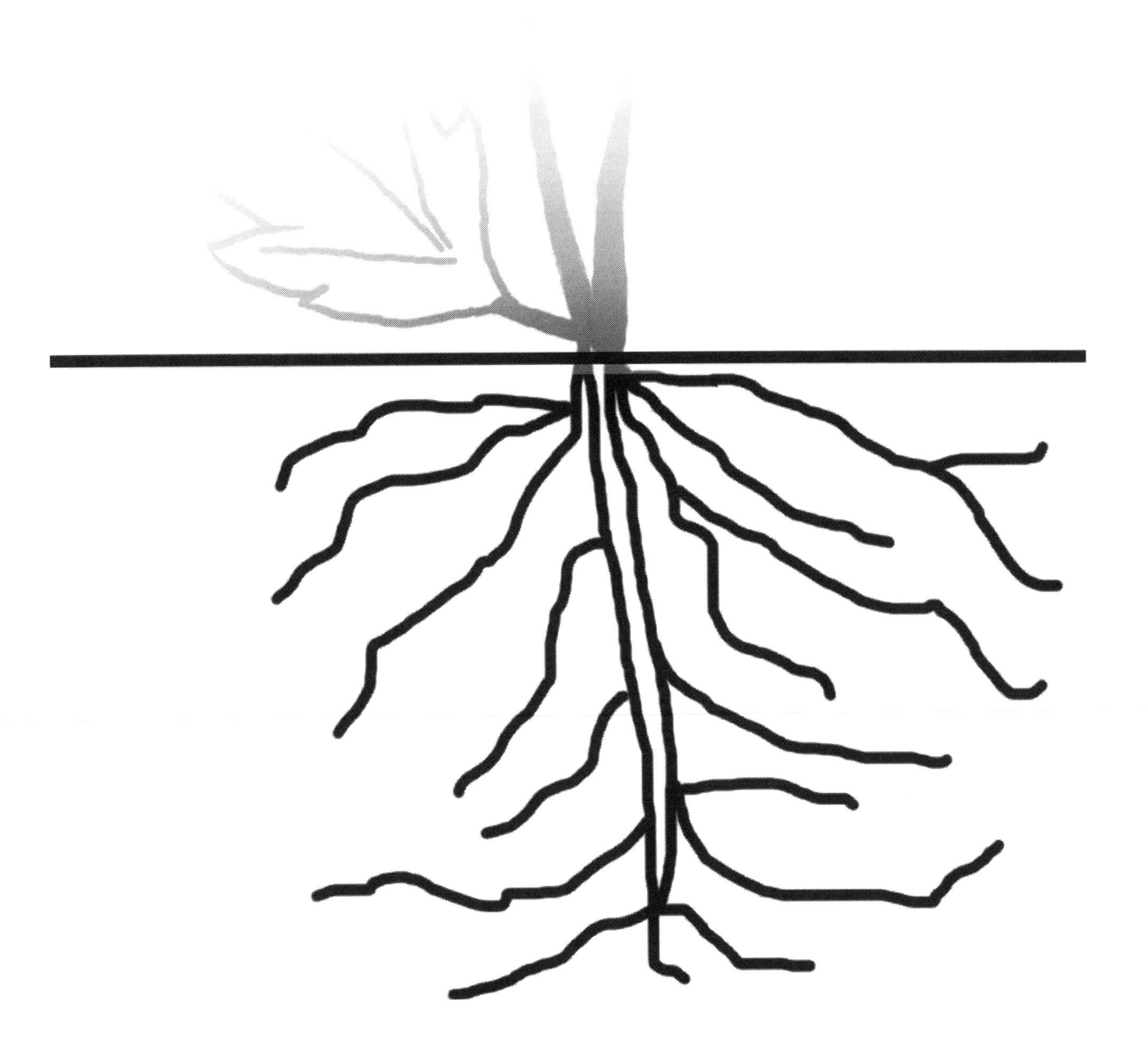

Roots take in water
and nutrients from the soil.

Growing Roots

What I learned:

Maple Tree

Painting with Roots

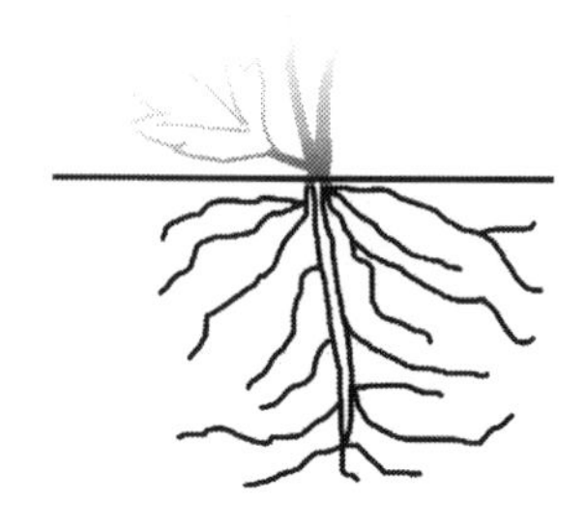

Intro to Science

Unit 6: Zoology Diary

Mammals, like rabbits, have fur or hair.

Comparing Mammals

Name of the Animal			
Size			
Hair			
Ear			
Teeth			
Food			

Rabbits

Mammal Collage

Reptiles, like snakes, are cold-blooded.

Cold-blooded

	Initial Temperature	Temperature after 2 minutes
Thermometer in the Sun		
Thermometer in the Shade		

What I learned:

____

____

____

____

____

____

____

Reptiles

Fingerprint Snakes

Birds have wings and feathers.

Cereal Feeder

Birds we saw:

Birds

Feather Painting

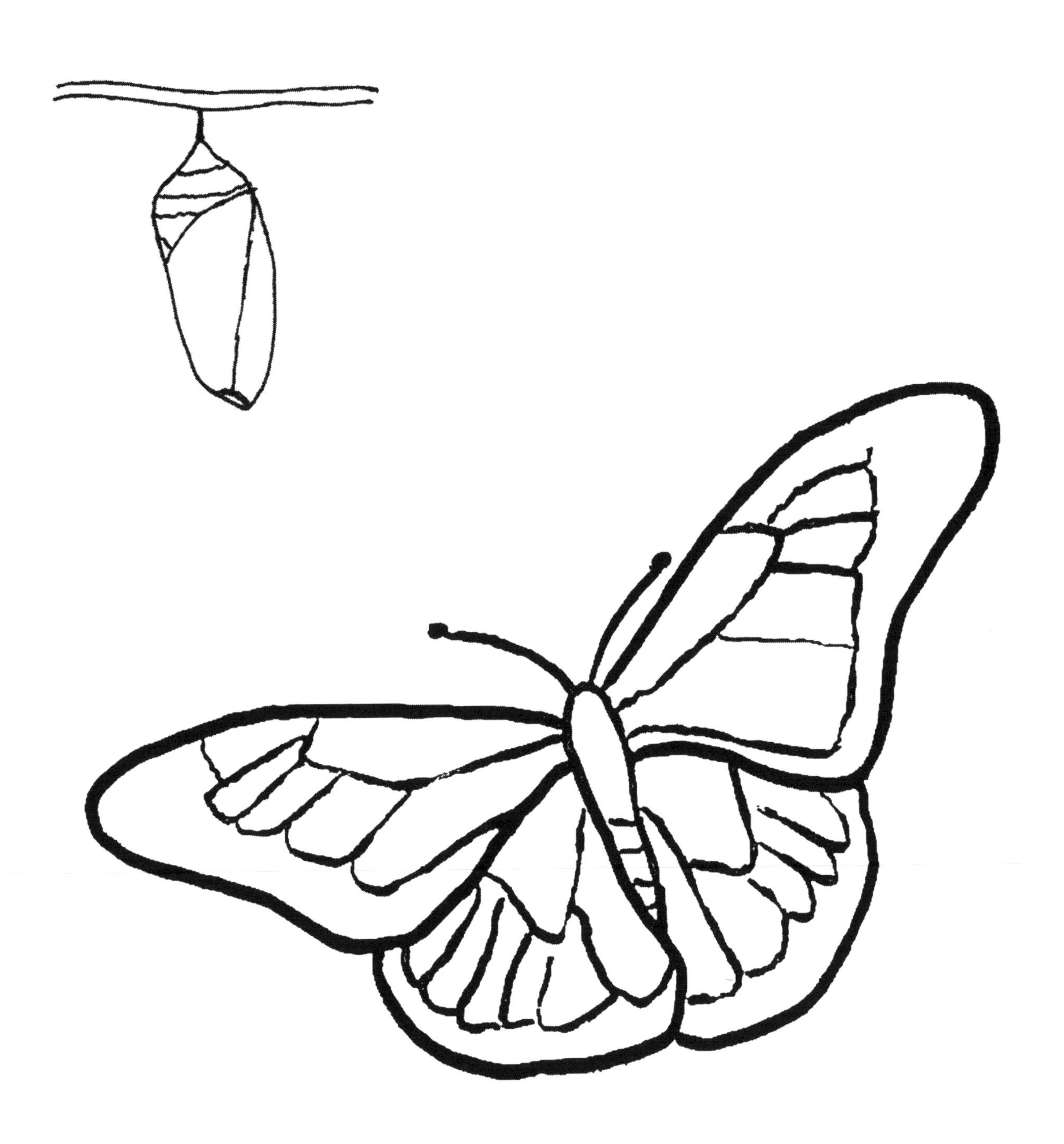

Caterpillars make a chrysalis
and then come out as a butterfly.

Butterfly Life Cycle

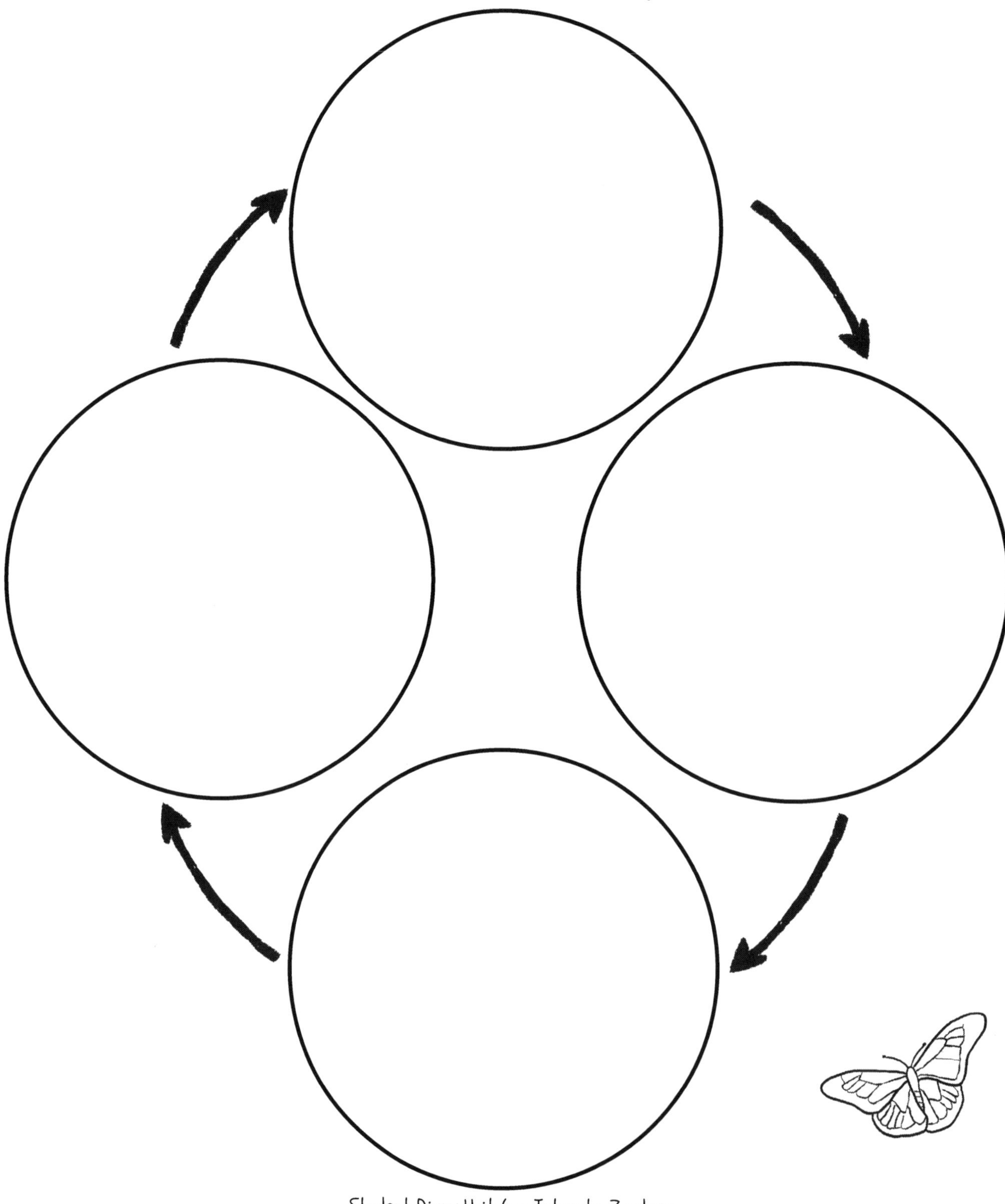

Butterflies

Butterfly Beauty

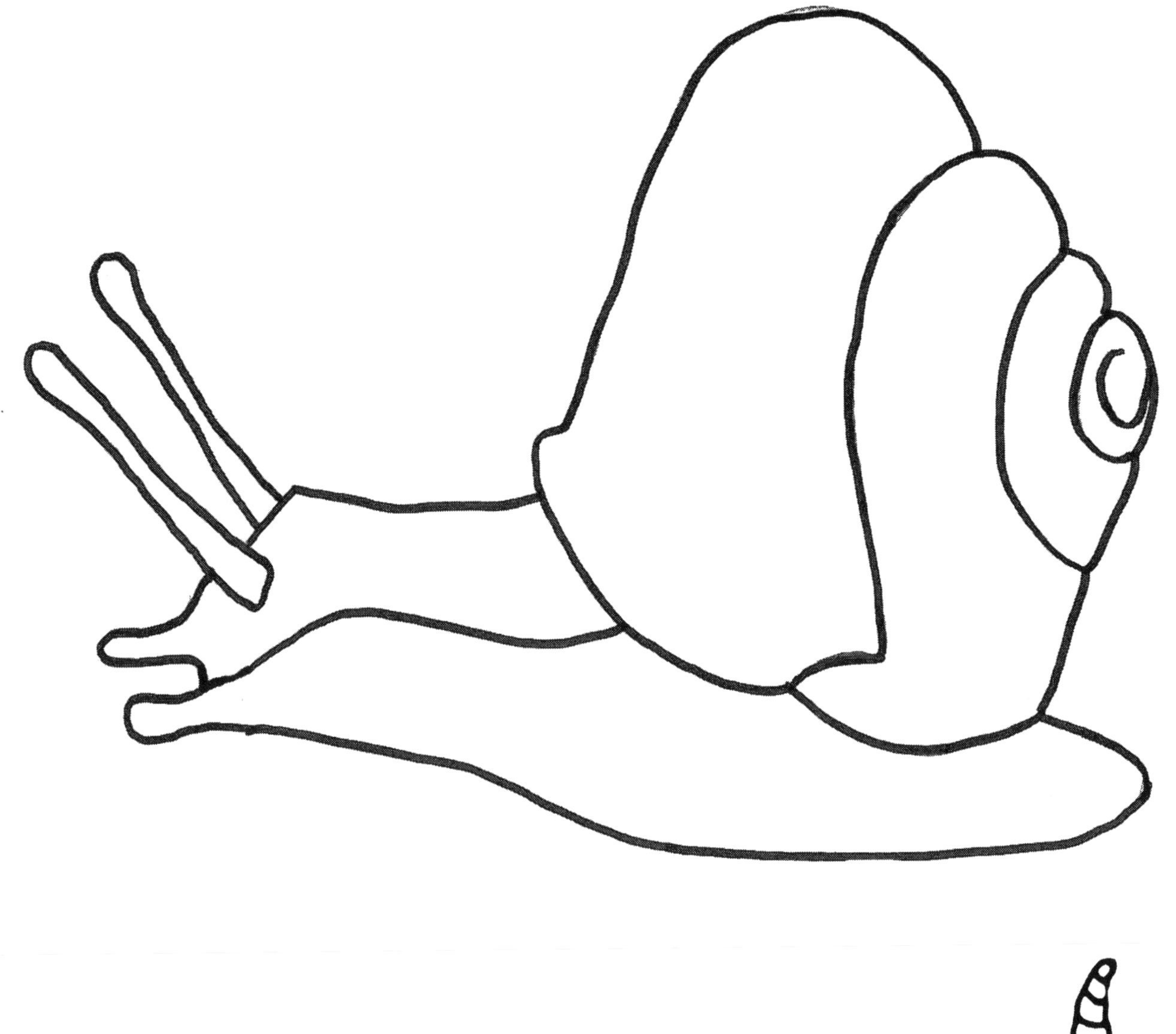

Invertebrates, like snails and worms, have no backbones.

Attracting Ants

What I learned:

Garden Snails

Worm Trails

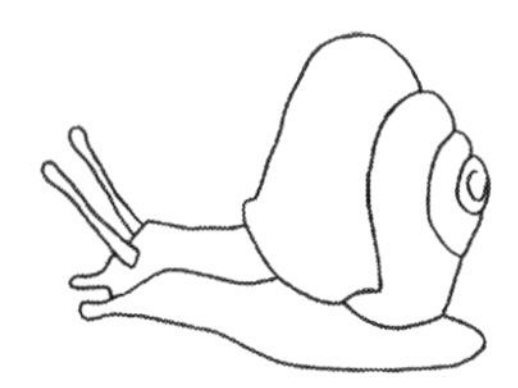

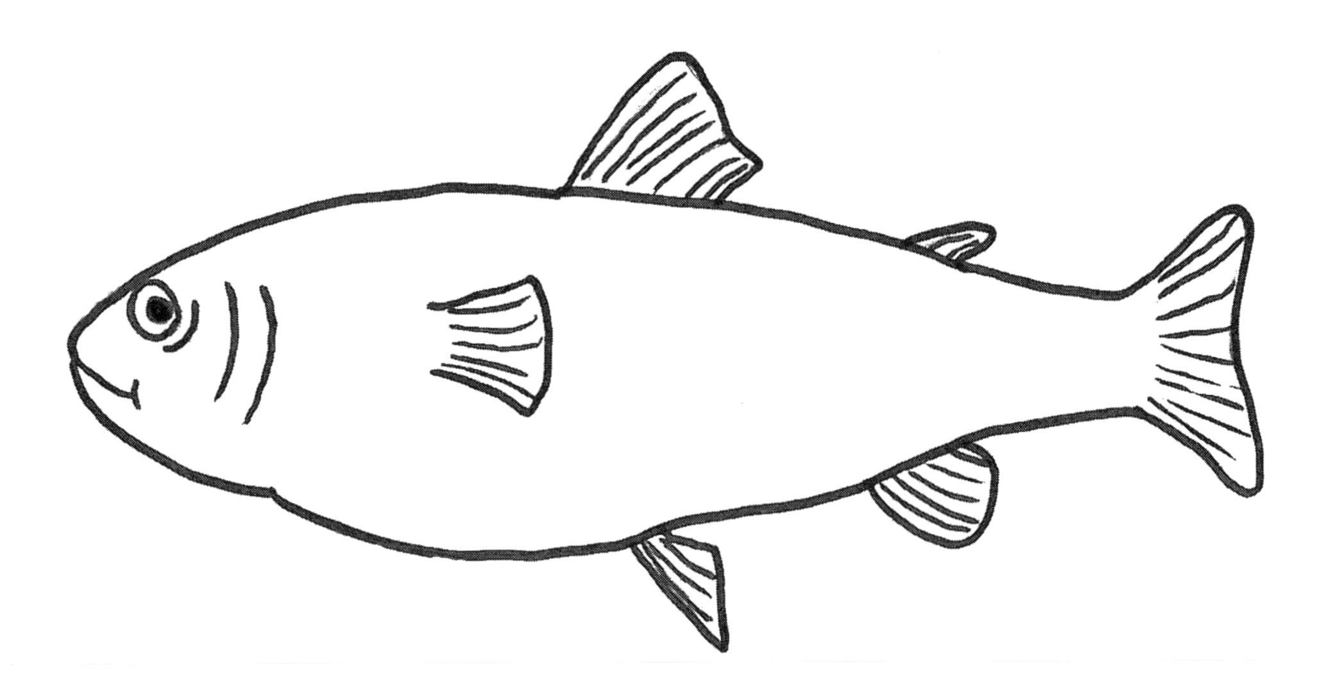

Fish have gills so they can breathe underwater.

Field Trip

My Favorite Animal

What I learned:

Fish

Sparkle Fish

Made in the USA
Columbia, SC
17 August 2024